James Heywood worked for fifty-eight years as a clogmaker in Fairclough's shop, Wigan. His father and grandfather were clogmakers before him. This photograph was taken in 1939, when he was seventy-one years old.

CLOGS AND CLOGMAKING

Jeremy Atkinson

Shire Publications Ltd

JOHN BIRD
Ground Floor
56 Leathwaite
London SW1
Tel: (071

JOHN BIRD
Ground Floor Flat
56 Leathwaite Road
London SW11 6RS-UK
Tel: (071) 228 5576

CONTENTS

History and use of clogs 3
Types of wood 12
Methods of construction 15
Sole styles 24
Irons 24
Crimping 27
Clogmaking today 29
Places to visit 32
Further reading 32
Clogmakers 32

Copyright © 1984 by Jeremy Atkinson. First published 1984. Shire Album 113. ISBN 0 85263 665 2.
All rights reserved. No part of this publication may be reproduced or transmitted in any form or by any means, electronic or mechanical, including photocopy, recording, or any information storage and retrieval system, without permission in writing from the publishers, Shire Publications Ltd, Cromwell House, Church Street, Princes Risborough, Aylesbury, Bucks HP17 9AJ, UK.

Set in 9 point Times and printed in Great Britain by C. I. Thomas & Sons (Haverfordwest) Ltd, Press Buildings, Merlins Bridge, Haverfordwest, Dyfed SA61 1XE.

ACKNOWLEDGEMENTS

The author expresses his thanks to Hywel Dafis, clogmaker, to the Reading Clog Dance Group, Howard Bamforth of the Colne Valley Museum, Judith Swann of Northampton Museums, Evelyn Vigeon of Ordsall Hall Museum and all those who contributed with help and information.

Illustrations are acknowledged as follows: Howard Bamforth, Colne Valley Museum, page 13 (both); BBC Hulton Picture Library, pages 1, 9 (lower); reproduced by permission of the Reference Library Department, Birmingham Public Libraries, page 7 (upper); Bradford Art Galleries and Museums, page 14 (lower); Mike Conquer, page 24; *Costume* magazine, Evelyn Vigeon and Salford Museum and Art Gallery, pages 26 (lower), 29; *Country Life*, page 10 (upper); D. R. Darton, page 2; Ailsa and Ian Dunmur, pages 25 (both), 31 (lower); Richard Halliwell, page 28 (both); Institute of Agricultural History and Museum of English Rural Life, University of Reading, pages 3, 5, 6 (lower), 7 (lower), 9 (upper), 12; Peter Murphy, page 14 (upper); reproduced by courtesy of the Trustees, The National Gallery, London, page 4; Chris Ridley, pages 26 (upper), 31 (upper); D. H. Robinson, page 10 (lower); Judith Swann, Northampton Museums and Art Gallery, pages 8 (both), 11; Al Vandenberg, pages 6 (upper), 15 (both), 16 (both), 17 (both), 18 (all), 19 (both), 20, 21 (both), 22 (both), 23, 27, 30 and front cover.

COVER: *Clogmaker Jeremy Atkinson, author of this book, at his workbench.*
BELOW: *The principal features of a clog.*

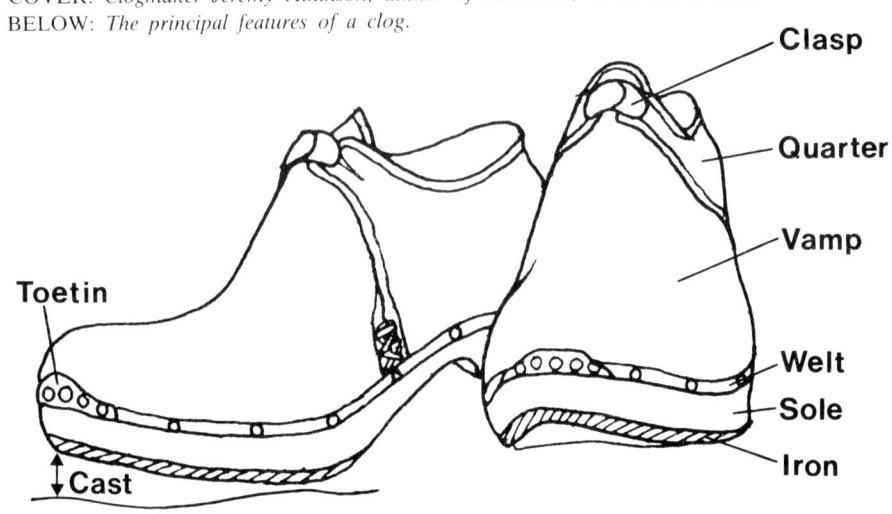

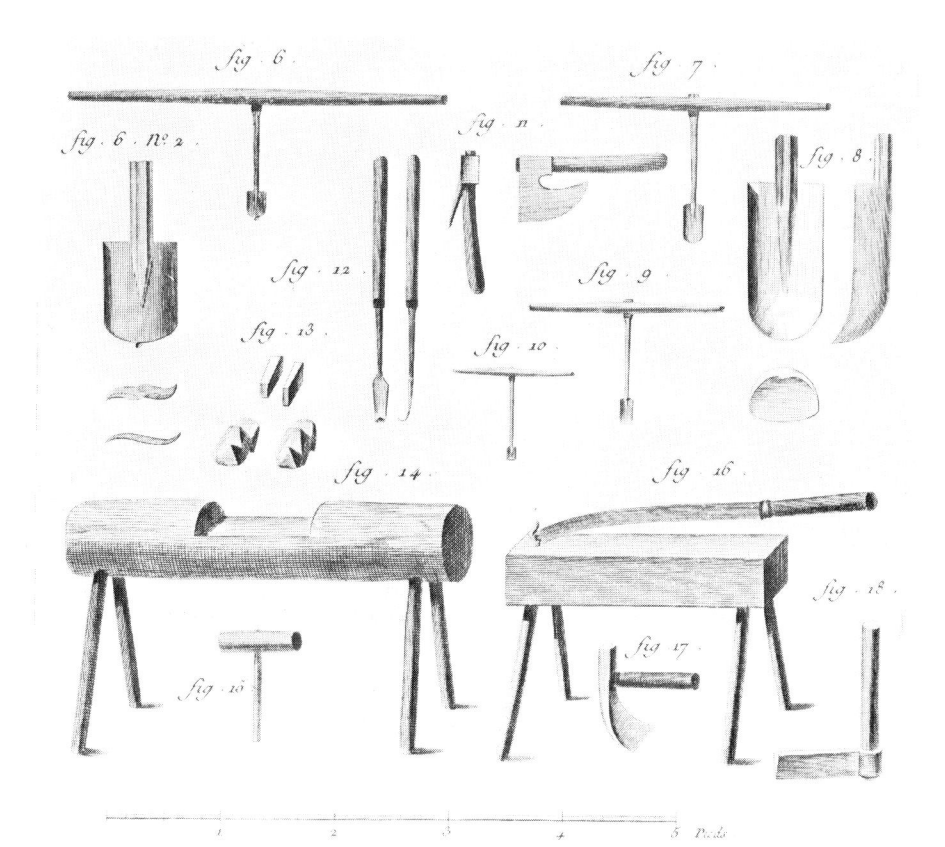

Tools used in the making of European sabots. The thin blocks shown above the sabots were used to wedge them in the bench while the craftsman gouged out the centre using both his arms and his chest. These tools have remained unchanged from medieval times to the twentieth century.

HISTORY AND USE OF CLOGS

Clogs have a relatively recent history in Britain. On the continent of Europe sabots (all-wood clogs) have remained unchanged for hundreds of years. Possibly because of England's break with the Roman Catholic church, continental influences have been less than might be expected, though in parts of Italy and northern Europe clogs strikingly similar to the British type are occasionally to be found.

In Europe the origin of all wooden-soled footwear is thought to be the Roman bath shoe, the purpose of which was to protect the wearer's feet from the hot tiled floors. From this derived the patten in its many various forms, and from the patten in turn the British clog originated. It is not known when the clog that we know today — that is the composite of wood and leather — finally developed. In Italy a pair of clogs with wooden soles and red leather uppers was found in the tenth-century tomb of King Bernard. In England the Wakefield Mysteries contain a reference 'His Luddokkys thai lowke like walk-mylne clogges'; and from the period 1600-2 there are references in the accounts of the Shuttleworth family from Gawthorpe Hall: 'clogging the boy Watmough a pr of showes' — which would seem to indicate clogs not pattens.

Although the most common continen-

'The Marriage of Giovanni Arnolfini and Giovanna Cenami' by Van Eyck, in the National Gallery, London. The pattens in the left foreground would be worn over stockings or simple moccasin shoes.

These pattens were owned by the vergeress of Winchcombe (Gloucestershire) and had been in her family since before 1750. Similar pattens were worn by the novelist Jane Austen, and people wore them to church in Montgomery (Powys) just before the First World War. Wearers were obliged to leave them outside the church!

tal clogs bear little resemblance to their British counterparts there is a strong similarity in their social status. All types of clog, whether British or continental, are essentially utilitarian forms of footwear, with little or no pretension to fashion.

This eventually led to their being shunned as a sign of poverty in the early twentieth century, even though shoes at the time were more expensive. Clogs cost perhaps a day or two's wages whereas strong shoes cost more than a week's wages in about 1900. Clogs were easily resoled, and so shoes were less durable, and they were no more comfortable. Nonetheless, clogs were looked down on, as people liked to ape their supposed betters, just as they chose white bread in preference to brown despite the greater nutritional value and superior taste of brown bread.

In the 1920s many Welsh people who habitually wore clogs during the week wore shoes or boots on Sundays. This was also common practice in many areas of England. In *Madame Bovary* the French writer Flaubert refers to clogs, once as being worn by a wet nurse and again as the footwear of a peasant woman.

Clogs are most suited to cold wet climates. They prevent heat from either escaping or intruding because of the poor conductivity of wood. They withstand water far more easily than all-leather footwear by raising the wearer's feet above the ground and because the wood itself absorbs less water than leather.

Clogs do have drawbacks, however. Because they do not bend, the wearer's feet tend to lift at the heel at the end of each stride, rubbing the sock against the leather. To counteract the rapid wear that ensued people would first dip the heels of the socks in tar and then in ashes to protect the wool. Nowadays, however, the widespread use of synthetic knitwear has reduced the effects of this problem.

Clogs are very unsuitable in wet snow, as the compacted snow sticks to them despite such traditional remedies as goose grease on the underside of the soles. Therefore the most common Scandinavian clog is a slip-on summer style. Clogs have had specific use in snow conditions: Captain Scott used alder clogs in conjunction with snow shoes on his ill

TOP: *This hinged patten was found in 1981 in the plaster of a shop at Kington (Hereford and Worcester). Another patten, identical save for being one size smaller, was found in the same year in the plaster of a house at Hay-on-Wye (Powys), 15 miles (24 km) away. The patten has a leather hinge with a leather sole studded with hob nails. Its date is unknown.*
ABOVE: *Made early in the twentieth century by George Besley, who was principally a wheelwright, this clog is similar to a type found in both Brittany and Holland, having a leather eyeleted strap over the arch only. It is shod with leather. Mr Warner of Llandrindod Wells made similar clogs between the wars. He was also a chairmaker and herbalist and he had an unusual use for alder cones: boiling these in water, he would make a cold 'tea' which when drunk regularly cured acne. Alder soles were often the preferred wear of northerners, who believed in the medicinal qualities of the wood for such ailments as rheumatism.*

fated final expedition to the South Pole.

Clogs were widespread in the British Isles by the mid nineteenth century though they have always been more popular in the north and west than elsewhere. In 1841 there were said to be 3246 clog and patten makers in England and Wales. Of these 1623 resided in Lancashire and in another year the proportion was almost two-thirds. Dorothy Hartley referred to clogs as 'rock shoes', which is particularly apt as they are less useful in heavy soils for the same reasons as applied to snow.

Clogs were to be found in all parts of Wales, especially in the south and west, where they were almost universally worn by country people and miners. In England they were most popular in Cumbria, Lancashire, Cheshire and Yorkshire and they were found in the Scottish border counties and on the fish wharves of the east coast of Scotland.

In the industrial north of England and

South Wales many different types of clogs evolved for different applications, but most commonly for use in wet cold working conditions. On fish wharves throughout Britain, from Billingsgate to Aberdeen, clogs of the *flap and buckle* types, often lined with felt, were worn to keep out the splashing water and the cold. In coal mines a special type of clog was used, known as a *blucher boot*. It was low-cut to enable a miner caught in a rockfall to slip out of his trapped clogs. Mill girls in cotton and woollen mills would wear simple *clasp clogs*, or in the twentieth century *strap clogs*. They often had two pairs so that one pair could dry out while the other was being worn.

As late as the Second World War clogs were still thought of as essential wear in many industries. A publicity pamphlet of that time lists fifty-three places of work where clogs were considered necessary.

RIGHT: *Thomas Butler, a Warwickshire peasant, wearing a pair of Derby clogs. This type of upper was common to both clogs and boots. Thomas James of Solva used the same pattens for both.*

BELOW: *Flap and buckle clogs, extensively used at laundries and fish wharves.*

The flap and buckle style (on the left) is derived from the blucher clog. Note the external leather heel support. The upper is of split kip and made by machine. The clasp clog (on the right) is made of better quality waxed kip; it is two-piece and hand-stitched. Both clogs have machine-made soles and are of twentieth-century origin.

DECLINE

Clogs began to decline in popularity around 1900. The industrial revolution, which had hastened the spread of clogs throughout Britain, also hastened their decline, for it brought about a rise in the earning power of the working man and at the same time reduced the cost of consumer durables through mechanisation and the economies of mass production. This enabled people to realise some of their aspirations. Leather shoes had become very much cheaper but even so, as late as the 1940s, clogs would have compared favourably in price with shoes of the same durability.

The First World War seems to have forced a brief resurgence in the trade but

A pair of 1930s child's one-piece button-bar clogs. The button is attached with either a pin or a staple. This pair is in the narrow round style with brass nails — there is no welt strip. Note the elegant flattening of the irons.

ABOVE: *There are few records of women clogmakers, though there have been some. Lack of strength to control the knives would have precluded most women from the trade, if social patterns of behaviour had not done so. These women are 'putting up', attaching the uppers to the soles. This is possibly a wartime photograph.*

LEFT: *An unemployed man and two children wearing clogs in Wigan, 1939.*

overall its effects probably accelerated the decline of clogs, weakening the social structures of the Victorian era.

In a Manchester newspaper article of 1930 it was stated that clogs had been in decline since 1865-70, when machinery began to be introduced in the making of boots and shoes; before this shoes and boots had been an unusual form of footwear, even on Sundays.

The survey of rural industries of 1927, relating to Wales, stated: 'The wearing of clogs has been decreasing for the last fifty years, and the decline is continuing... The competition of machine-made clog soles is keeping down the prices. In Lancashire factories are swamping the market with beech clog blocks and it is thought that as the machinery is im-

ABOVE: *Splitting blocks of wood for clogmaking, 1900.*
BELOW: *A clog blocker at work in Minsterley, Shropshire, near the Welsh border, in 1940. The block is held against an old file on the bench for grip. The knife is prevented from touching this by the piece of wood set across the bench at the opposite end from the ring.*

proved the demand for Welsh clog blocks will cease. The number of apprentices to the trade is decreasing so fast that there was hardly one found in the whole area investigated, most of the cutters being past the prime of life.'

Mrs Turner, who lived in Gladestry in Powys on the Welsh marches from before the First World War to 1982, had a brother-in-law who was a master clogger, that is a clogmaker skilled in all areas of the craft, both wood carving and leather work, and her husband worked as a clog sole blocker in the alder groves which still abound along the streams of this area. Before the First World War her husband made a reasonable living as a blocker, supporting his family solely from this source of income. He returned wounded from the war and took up his old occupation. However, by the early 1920s the blocker gang found the work no longer paid. The price dropped with demand.

Mrs Turner recalled that the last stacks of blocks were used as firewood and that her husband then went to work in the local quarries. This seems to confirm the government report.

However, there were still clogmakers working in many areas until the Second World War and through it, when the government encouraged people to wear clogs by a vigorous publicity campaign and by allowing people to buy clogs without coupons, which were needed for shoes. At Carmarthen, for instance, there were at least half a dozen clogmakers selling from the market during the war and it is likely that some of these were master cloggers making the entire article, because many west Welsh people preferred hand-cut soles, which in south-west Wales were made of sycamore, to factory-supplied soles, which had formerly been made of alder although by this time they were being made from beech.

This clog was made in the Second World War by a shoe manufacturer. The wood is ash, traditionally used in dance clogs for its clarity of sound. In this case the sole has been steam-bent, with a heel piece glued in, and every effort has been made to disguise that it is a clog. Other clogs made by shoe manufacturers during the war were sometimes hinged at the sole.

A clog-blocking gang posing for the camera in 1910. The rough-cut blocks, known in the north of England as writhings, were stacked in this way to allow air to circulate and dry them. The wood being blocked here is birch.

TYPES OF WOOD

There is still much argument over which is the best soling material. Old cloggers in Yorkshire invariably praise the merits of alder. Alder was the most popular wood in the pre-mechanised era but it was by no means universally preferred. It has the advantages of being light, of being easy to cut both dry and green and of taking a nail easily when dry without splitting. It absorbs sweat well in hot factory conditions and in towns seems to last well. Its disadvantages are that it is soft when dry and that it is quickly eaten away by acidic soil conditions — not a problem in the limestone areas of the north. Furthermore, if left damp and muddy it rots quickly.

In the survey of 1927 Caernarvonshire clogmakers stated 'that the reason of the extensive use of alder elsewhere is that it is of no value for any other industry and hence is cheap'. In support of this, one man stated that sycamore cost him one shilling per foot (0.3 m) whereas alder cost only three pence per foot.

Birch was the preferred wood of a Cumbrian clogmaker, Mr Carruthers of Cleator Moor. His father had been a master clogmaker and together they would go to Scotland by lorry to select and cut their birch trees. They would bring these back and cut them into large sections, leaving them to dry out for at least three months before they were split into clog blocks and carved. They made clogs for the local mines and found that alder broke up in the acid conditions. Birch, which contains resin, was more resistant.

They used sycamore only for women's clogs, possibly because they worked it in the same way as birch and were perhaps deterred by its hardness. Miners in Flintshire also demanded birch soles, which were brought in from clog sole mills such as Maude's of Hebden Bridge, so it is possible that birch resists acidic water better than sycamore.

Sycamore was the favoured wood of west Wales and interestingly this is the only area that had an abundance of all three types of wood. The local population preferred sycamore for its durability. The 1927 government survey states: 'In the Teifi valley in south Cardiganshire and north Pembrokeshire every village has a clogmaker who makes complete clogs, and here not only the farm servants but also the women and children in the villages wear them. These people will have no other but sycamore, and believe that sycamore is more comfortable for the feet and more durable, being closer grained than alder.'

One of the last remaining master clogmakers in Wales was Thomas James of Solva, Pembrokeshire (Dyfed). He would use nothing but sycamore if he could help it and worked sycamore in a different way from English clogmakers who used alder and birch. He would cut trees only in January and he worked the wood green, letting the wood dry out on the wearer's foot. This is the key to the Welsh use of sycamore, as when dry it is too hard to be worked commercially by the use of traditional clog knives. Probably the Welsh cloggers found out how to work sycamore from the Welsh turners. In the Teifi valley there was a thriving turning industry based on sycamore. One of the methods used was to cut a nest of bowls, the outside of one being the inside of the next. This was done with green sycamore and the turners' knowledge of the ways of working sycamore must have been imparted to the clogmakers, who would have lived in the same villages up and down the Teifi.

Thomas James stated that hedge-cut sycamore was superior to that from a copse. This may have been due to the older root structure of the hedge sycamore.

Much of the wood for the north of England came from the Welsh marches: alder from that area was considered to be of the best quality though clog blockers also worked in other parts of Britain and even in Ireland. Clog blockers such as Mr Turner would send their piles of rough-cut blocks to the north by railway. The blocks were sized as men's, women's and children's and at their destination they would either be finished in a clog sole mill and then sent out to clogmakers to have the uppers attached, or be sent on in their unfinished state to be hand-cut to their own satisfaction by the individual clogmakers. Both alder and birch cut better dry than wet and so were ideally suited to this practice.

ABOVE: *A clogger's workshop of about 1910 at the Colne Valley Museum near Huddersfield.*

BELOW: *James Horsfield made both brass and japanned steel clasps and buttons for clogs in a wide variety of sizes and styles.*

ESTABLISHED 1887.

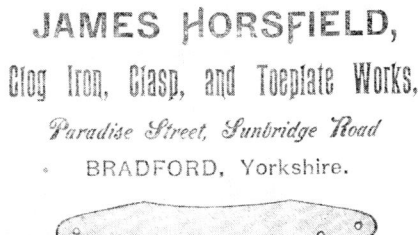

JAMES HORSFIELD,

Clog Iron, Clasp, and Toeplate Works,

Paradise Street, Sunbridge Road

BRADFORD, Yorkshire.

Illustrated Lists of Clasps and Toeplates on Application.'

I ALWAYS have in Stock a large and varied assortment of Clog Irons suitable for any District, and can therefore supply almost any shape and quantity, without delay. My Clasps and Toeplates are made from the very best materials, and are second to none for durability and finish. Agent for the best makes of Clog, Tip, and Brass Nails, and also for Henry Carter's Knives.

By the inter-war period factories were buying whole trees and cutting their own soles from start to finish. At first they used sycamore but later changed to beech, because the silica in sycamore blunted machine blades.

Beech, being harder than sycamore, was a wood little favoured by hand carvers. All factory soles are now made from beech. It is very durable, easily cut across the grain because of the shortness of the fibres and extremely hard when dry — harder than sycamore. On the other hand, it has little or no spring, having a dead feel; and being short-grained it is inclined to split across the grain when the soles are worn thin. It is also the heaviest of the clog woods.

Other woods that were used locally were willow (in Cheshire) and poplar. Both these are very light and poplar is very resistant to abrasion. Poplar might have been used more had it not been for its scarcity. This prompted Henry III to pass an act forbidding its use by clog and patten makers because of the demand for it in the making of arrows and bow staves. Edward IV passed a similar act two hundred years later. After both acts there were petitions on behalf of the clog and patten makers for the rescinding of the act: 'It is soo, right-worshipful sir, that the seid Tymber of Aspe is the best and lightest Tymber to make of Patyns and Clogges.'

ABOVE: *Hywel Dafis of Tregaron (Dyfed), one of the few clogmakers practising today. He advocates the cutting of alder at all seasons, roughly shaping the blocks and leaving them to dry slowly. The belief that alder should be cut only in spring and summer arose because it grows in wetlands, where it is easier to work in the drier seasons.*
BELOW: *A clogger's sign. Displaying such a sign was common practice in England.*

LEFT: *Splitting the block for a matched pair of soles.*
RIGHT: *Levelling the underside of the sole with the stock knife.*

METHODS OF CONSTRUCTION

SOLE CARVING

By 1980 there were only three or four practising hand carvers left in Britain. Carving is much the most difficult skill to master, and it was reckoned to take ten years to be fully skilled. Carving is difficult because it requires a combination of strength and control and is largely done by eye and feel. Each sole has to be a mirror image of its pair. In the past a master clogmaker would measure customers by simply feeling their feet with his hands and he would then transfer this straight on to the block of wood, using his experience and knowledge to judge the size. Thomas James, when asked what he used for a pattern to guide his hands, picked up an old sole and said 'this': he would scale it up and down by eye. That sort of skill has gone.

Three clog knives are used; these are attached to a ring in the bench and, with the block of wood held in the left hand against the left thigh, the clogmaker levers the blade down on to the block with his right hand.

The first knife used is the *blocker*, or *stock knife*, which has a straight blade not unlike that of a butcher's knife. This is used to cut the base and sides of the sole and roughly to shape the inside.

The second knife is the *hollower*, which is used to hollow out the sole for the foot to rest on. Both the blocker and the hollower are made in two pieces, which are welded together. The blades are of hard carbon steel.

The last knife used is called the *gripper bit* or *rebate knife*. This cuts the groove for the upper to sit in. This is the most difficult knife to use as the carver is following the curve made by the previous two knives and having to move both the block of wood and the blade in tandem in one flowing movement. The blade of the gripper bit is bolted to the arm. Some

ABOVE: *Roughing out the shape — always working towards the centre of the grain.*

BELOW: *Carving the 'cast', the curved underside of the sole.*

ABOVE: *Cutting the heel. Alder is soft enough for this to be done with the stock knife, but sycamore is not.*

BELOW: *Trimming the heel.*

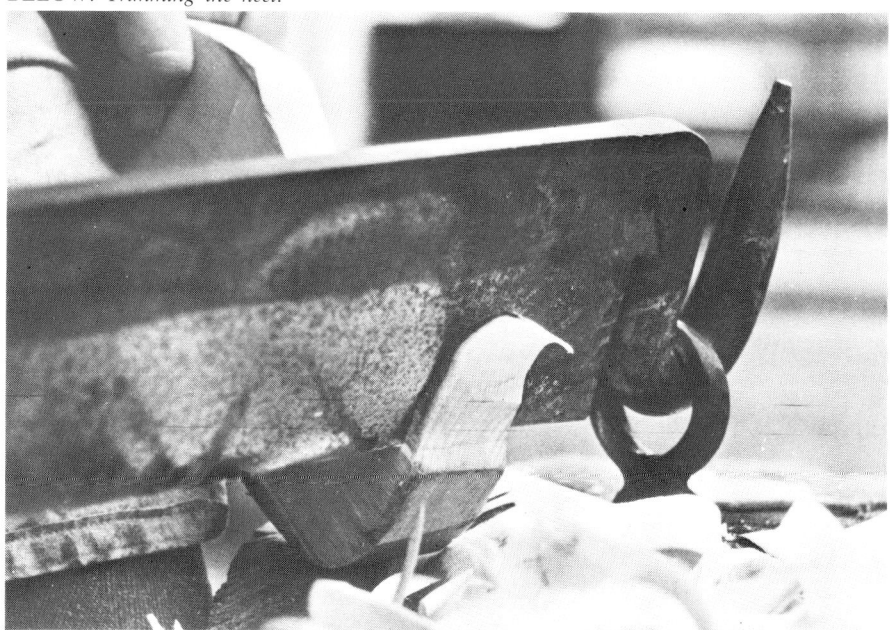

ABOVE LEFT: *Almost as much as can be done using the stock knife.*
ABOVE RIGHT: *Hollowing out the sole.*
LEFT: *Trimming the edge in preparation for the gripper.*

knives had more than one slot to enable the clogmaker to choose where he put the blade along the arm.

In working the knives there are two golden rules: the knives must be razor sharp, and the craftsman should always work towards the centre of the grain.

UPPERS

Although there is argument over the best type of wood for clogs, opinion is united over what was the best leather. This was waxed kip and was about ⅛ inch (3 mm) thick, black, and with the flesh side out. It is unfortunately no longer available. The whole surface was coated in wax, which permeated the leather, making it exceptionally supple when new, and when old it was like iron. The waxed surface was so heavily treated that it shone.

The leather would be cut with a clicker knife and stitched with a butt stitch that drew the leather together edge to edge in

ABOVE: *The gripper bit.*

BELOW: *Cutting out the upper.*

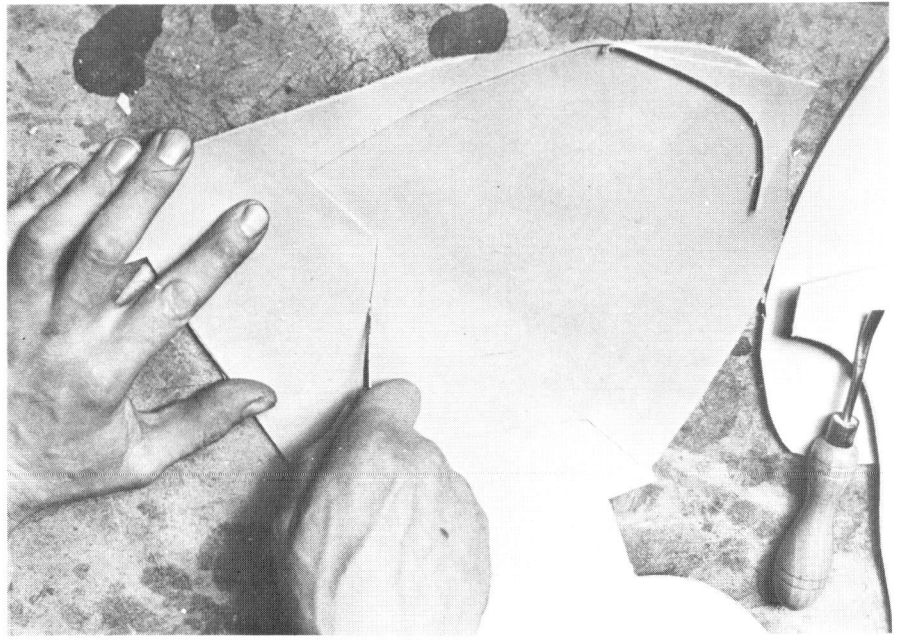

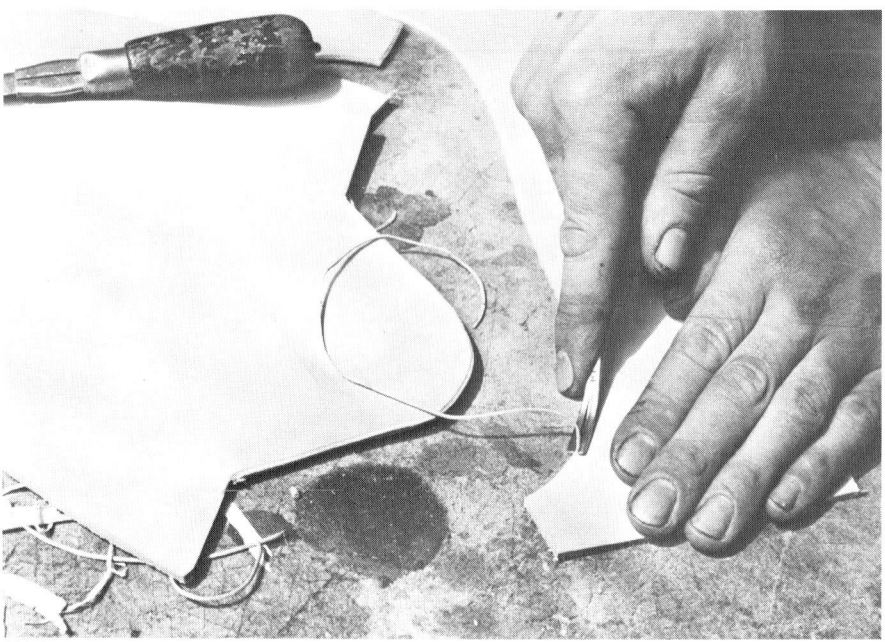

Edging.

an operation called *closing*. This type of stitch was ideally suited to wax-kip leather, which had a very high fibre strength. It was also the stitch used with the oak bark leather that preceded kip.

The two pieces of leather were placed on a horizontal pole of about 3-4 inches (80-100 mm) diameter that was held in place by a loop in the bench. A further strap of leather was placed over the two pieces of leather and held in place by the stitcher's foot, leaving both his hands free to stitch. The stitcher would commence to stitch by making a hole in the face of the leather about ⅙ inch (4 mm) back from the edge, angling the curved blade so that it protruded from the edge of the first piece of leather and entered the edge of the second piece about two-thirds of the way down from the waxed outer side. The awl would then be driven further in, finally making a hole ⅙ inch (4 mm) back from the edge of the second piece. The awl would be withdrawn and the stitcher would simultaneously push in thread from both sides so that the bristles slide past each other in opposite directions.

The thread, being whipped to the bristles, would be pulled through and then the stitcher would make a second hole about ⅙ inch (4 mm) further along the edge and repeat the process. To finish, the stitcher would repeat the process until the two pieces were 'closed' and then go back three holes and cut off the surplus thread. The thread itself was hand-made by the clogmaker from six strands of cotton, tapered, twisted, waxed and whipped to the boars' bristles. The bristles were imported from Russia.

Later, people would either buy ready-made traditional one- and two-piece tops or machine-stitch their own simply by overlapping the leather.

Thomas James, machine-stitching his own tops and cutting his own soles, regularly turned out four pairs a day at his peak, taking something under an hour to cut his own soles.

By the inter-war period multi-piece uppers were coming into use and became so widespread that they are now considered traditional, having almost entirely superseded the old styles which were

ABOVE: *Closing.*

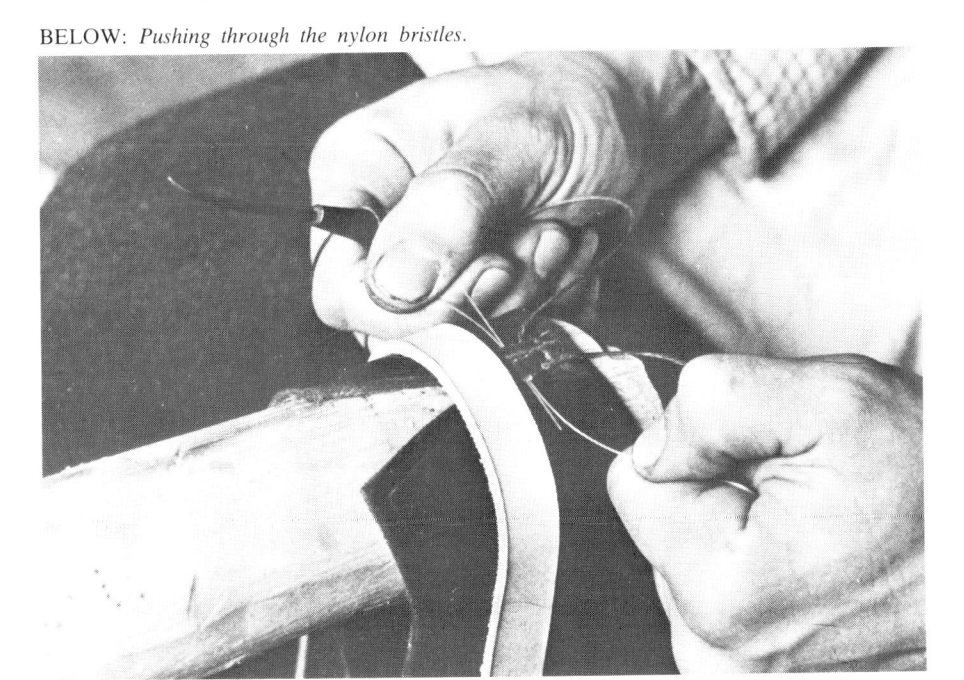

BELOW: *Pushing through the nylon bristles.*

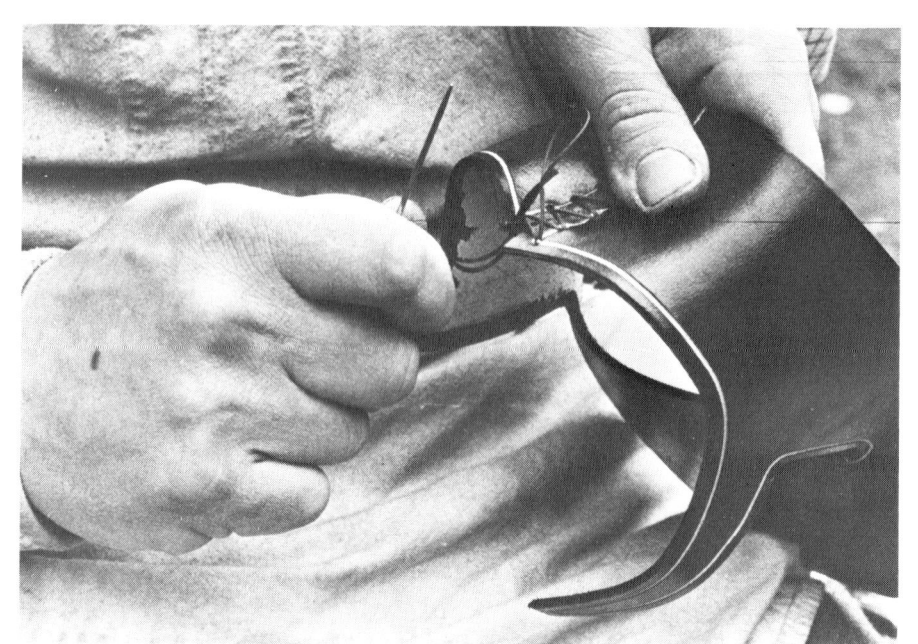

ABOVE: *An alternative method of stitching the uppers, using modern braided synthetic thread instead of bristles.*

BELOW: *Lasting — stretching the leather over the last with lasting pincers.*

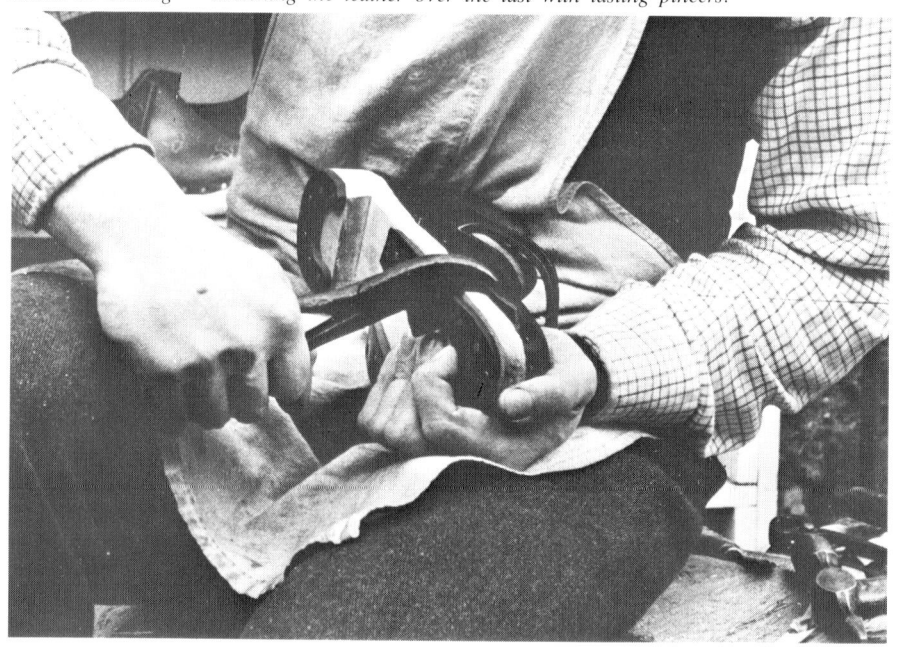

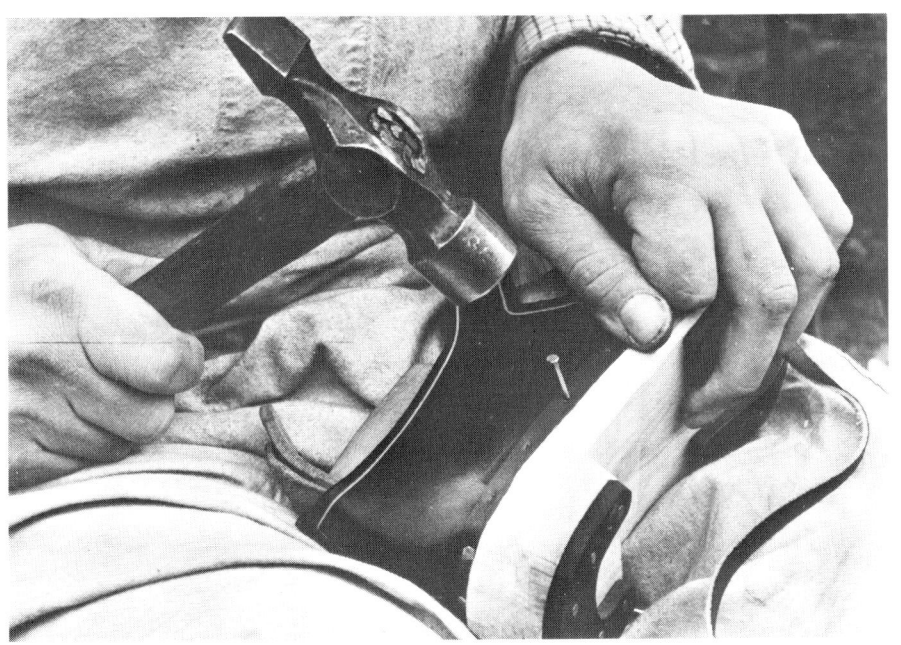

Nailing on the welt strip.

evolved with hand-stitching in mind.

LASTING

The methods of attaching the upper to the sole are straightforward, but before this is done the leather has to be shaped. Originally the lasts were 'straights' and solid. The leather upper was placed over the last, which the clogmaker held on his legs.

With waxed kip the method was to heat a glazer with a half-round bottom over a fire or gas flame and then rub the glazer in a sliding motion over the leather, melting the wax and allowing the leather to be pulled down tight over the last with lasting pincers. The upper would then be attached to the last by means of light tacks nailed to the underside through the surplus leather. The wax having reset, the leather was taken off the last and placed on the sole. The inner edge was twisted slightly forward of the outer edge to provide more room for the ball of the foot and the upper was nailed on to the rebate of the sole, the join often being covered by a strip of leather called the welt. Toetins would be attached to protect the toes, and the clasps or buttons, depending on style, would then be attached.

Later, paired sprung lasts were used and the upper would be nailed on to the sole with the last in place. The clogmaker would use short nails ⅜ inch (10 mm) long at the toe and gradually increase the size towards the heel to ⅞ inch (23 mm) or 1¼ inches (30 mm), depending on the size of clog.

With waxed kip the lasts could be taken out as soon as the wax was cold, which would be well before the clogmaker had finished attaching the welt and toe plates. However, leather treated with oak bark or vegetable tan is soaked instead of heated and the uppers must be left to dry before the lasts are removed.

A hand-made pair of clasp clogs with elegant square toes, hand-stitched kip leather, hand-cut soles and no irons; these almost certainly date from the nineteenth century.

SOLE STYLES

Not only were there different types of upper for different uses, but there were also different styles of soles, known as *ducks, pins, common rounds* and *London*, with many variations within these main styles.

Ducks were worn in many areas of the north and their strange shape is more practical than it might at first seem. The extended toe allows for more *cast*, or curve, to be built into the sole, thus making it easier to walk up the steep sides of the mill valleys of the Pennines. Pins were similar but even more pointed and were popular in Scotland, amongst other places. The London style was otherwise known as *square toe* and this shape was especially useful in industries such as mining where a lot of work was done whilst kneeling, so causing a great deal of wear on the toes. Steel workers also wore square-toed clogs; with this shape, it was easier to cover the front of the foot with a protective steel toe plate. The London style also had a much squarer and more stable heel.

Common round was the normal style throughout Wales.

As well as making different styles of clog soles, the factories also made them in different thicknesses.

IRONS

An iron was traditionally attached to the bottom of the sole. In very poor areas one could buy clogs without irons, or *calkers*, attached but this was a false economy as the clog's life would then be much less than that of an ironed pair.

Irons were originally made by hand. The clogmaker would either buy irons

ABOVE: *A child today with feet that would fit into these clogs would still be crawling. They are worn down at the heel. The inscription stamped on the upper reads 'C. R. Rowther, Clogger, Burnley'.*

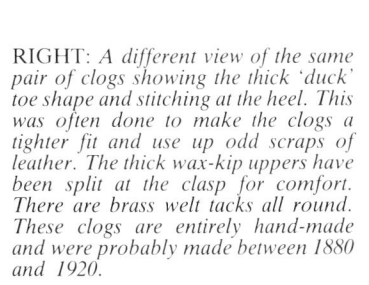

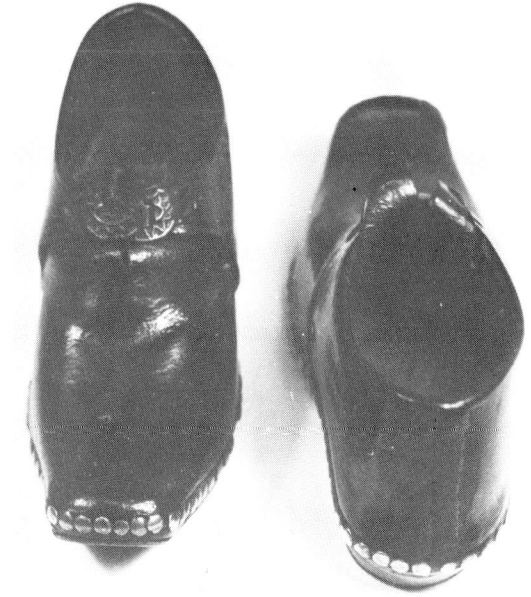

RIGHT: *A different view of the same pair of clogs showing the thick 'duck' toe shape and stitching at the heel. This was often done to make the clogs a tighter fit and use up odd scraps of leather. The thick wax-kip uppers have been split at the clasp for comfort. There are brass welt tacks all round. These clogs are entirely hand-made and were probably made between 1880 and 1920.*

ABOVE: *A child's two-bar clog, another one-piece pattern. This clog is thought to date from the mid nineteenth century and is made of waxed kip leather decorated with a simple crimp.*
BELOW: *Clog irons. (From left to right, top) Thick duck, otherwise known as duckbill or duck neb. Colne strong irons, which were said to last longer. (Centre) Common round. Narrow round. (Bottom) Square half-round. Women's square round. (From 'Costume' magazine.)*

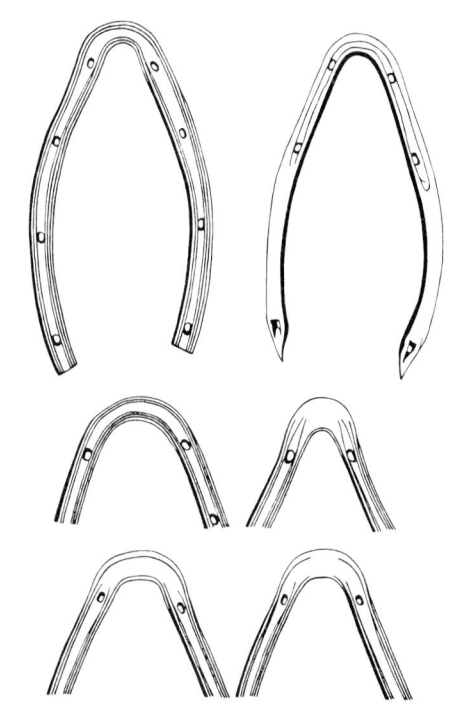

from the blacksmith or send his customers to the blacksmith to have a pair specially made.

Originally the irons were not attached to the underside of the sole but to the side of the sole. These were called ring clogs. An early example is the Etruscan wooden sole bound by a bronze hoop in the Bally Shoe Museum in Switzerland.

Irons similar to those still used in steelworks perhaps came into use in the early nineteenth century but it is likely that the two types co-existed for many years.

The manufacturing of irons by machinery originated in the late eighteenth century; first a production line technique was used and then machinery was introduced. This machinery was also used in north-west France, where a similar leather and wood composite clog is still worn today, normally over felt slippers. The irons were made in as many variations as the soles.

When a pair of irons had worn out the wooden sole would have new irons put on, the clogmaker filling the old nail holes with oak pegs.

A crimped narrow-duck clasp pair and a crimped common-round button-bar pair made by the author and crimped by Pat Kent.

CRIMPING

Although clogs have always been essentially practical they were not always plain. Apart from the attractive shaping of the soles and uppers there was also an art form peculiar to clogmaking called crimping. Crimped clogs were worn for Sunday best whereas work clogs were normally left plain. There were also strict rules of etiquette about what sort of welt tacks were nailed into the rebate of the clog. In certain areas only women had brass tacks and toetins. Most of these unwritten rules seem to have applied only in the north of England, where the art of decoration was taken further than in Wales.

Crimping involved cutting patterns in the leather by the use of a tool called a *race*. Each clogmaker had his own patterns. Flowers, butterflies, fish, stylised forms of these and other designs were used to adorn the clogs, together with geometric patterns, often in the form of cross hatching. Clogmakers were often extremely jealous of their designs and it was not done to copy the patterns of others.

This tendency probably increased during the twentieth century. Once a craftsman's clogs could be recognised by the way the sole and the upper were cut, so that one could tell where a person lived by his or her clogs. Later, however, as clogmakers first bought in finished soles and then finished uppers, there would be little to tell one man's work from another's, except his crimping. There would be little skill in anything apart from the crimping. This tendency towards standardisation happened so quickly that most cloggers coming to the end of their careers in the 1970s and 1980s had never used clog knives.

ABOVE: *Hand-made crimped blucher wax-kip dance clogs, thought to be of nineteenth-century origin.*

BELOW: *These dance clogs were made in 1981 by John the Fish of Truro, Cornwall, who is one of the few present-day clogmakers. They have hand-made vegetable-tanned uppers and are multi-coloured, with modern leather stamps as well as traditional crimping. They have hand-altered beech Walkley soles with the edges rounded out to protect the leather during dancing. Conventional lasts were not used, the clogs being made to fit each foot.*

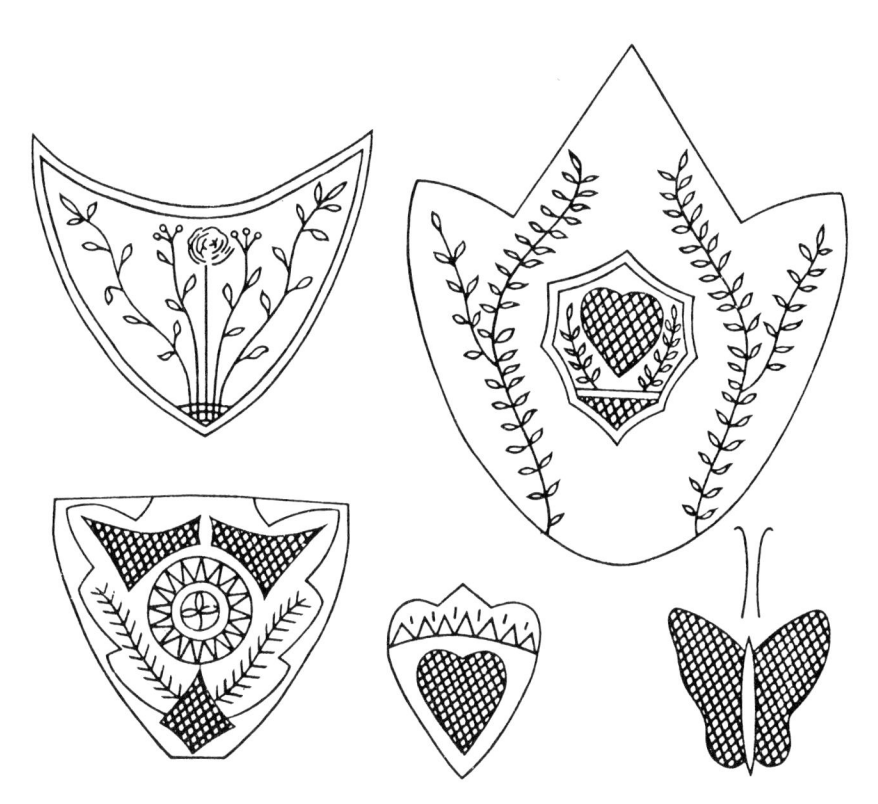

Traditional crimping designs (from 'Costume' magazine). (From left to right, top) By Harry Bell of Salford, 1966, in Salford Museum and Art Gallery. Crimp design held by Salford Museum and Art Gallery. (Bottom) Crimp pattern by Mr Brew of Whitehaven. Two alternative centre designs by Mr Brew. Although Mr Brew's career started well before the Second World War he confided that he had never used his clog carving tools to make soles — always buying them ready made.

CLOGMAKING TODAY

Today only the factories and a very few individual clogmakers cater for industry. The National Coal Board ceased using clogs after the Second World War, changing to protective boots. With the shrinkage in the steel industry demand there too has fallen.

To counter this, Walkley's, who took over Maude's clog sole mill at Hebden Bridge, have expanded their leisure lines.

Older craftsmen have been slow to adjust to the changing market. As shoes have become less expensive since the Second World War clogmakers responded by buying in cheap ready-made soles and uppers. This eventually led to a uniformity that was not helpful in increasing business.

Cloggers in the industrial areas of the north often bought old shoes from better-off areas and used these as uppers on clog soles to hold down the price. This was common practice in the inter-war period but now the economics are far less favourable than they once were.

Clog dancing is the cousin of American

General clogging tools. (Clockwise from top right) Stiddy, half-round bottom glazer, edger, nail extractor, clicker, Stanley knife, lasting hammer, pincers, bradawl, race, scraper, leather soling knife, lasting pincers, button pin gun, stitching awl, button hole punch.

step dancing. Many of the steps are remarkably similar although the footwear is not.

In Britain there are many different local styles of clog dancing. The Durham style has more heel beats than is usual and so there is a specific Durham dance clog which has very little cast in the sole to allow the dancer to come off the toes more rapidly to accommodate the heel beats. Black Americans added syncopation to step dancing to make tap dancing, and so clog dancing could be said to be the ancestor of tap.

The revival of traditional music and its accompanying dances and traditions has provided a growing market for clogmakers. Many clog dancers take great pride in their kit and readily pay far more in order to have clogs custom built. Younger craftsmen less inhibited by the association of clogs with poverty have been quick to exploit this.

Thus although there were only three or four people in 1980 capable of cutting their own soles there were others who would alter Walkley's soles to their own satisfaction. More craftsmen also began to make their own uppers again, both in modern machine-stitched multi-piece styles and in the older one-, two- or three-piece hand-stitched style. People began to innovate, with variations on both upper design and styles of crimping.

By 1983 there were nearly forty practising clogmakers left in Britain, many of these young newcomers to the trade. Of these two or three would cut soles, four or five would alter factory soles to suit, roughly fifteen would make their own uppers and perhaps twenty-two would crimp. But this is an enormous reduction from the 6276 recorded for England and Wales in 1901!

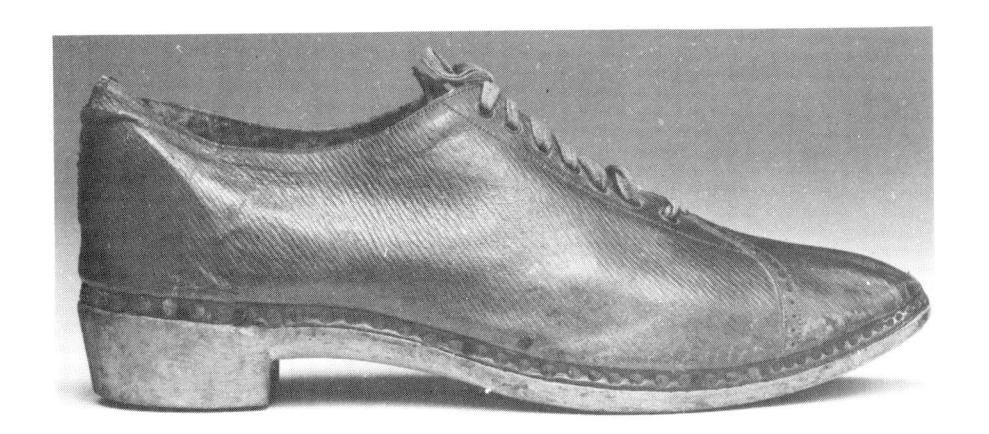

ABOVE: *A red dancing clog with black welt and blue laces, made in the First World War for Lieutenant Frank Charlton. There is very little cast to the sole and the heel is hollow — a penny was put inside to rattle. The uppers are more like shoe uppers than traditional clog uppers and were made of patent leather. The soles appear to be made by hand.*

BELOW: *Cumbrian dance clogs that belonged to Ailsa Dunmur in the 1930s. The toes are of dark red patent leather with brass tacks, the heels of dark blue patent leather with steel tacks.*

PLACES TO VISIT

Intending visitors are advised to find out the times of opening before making a special journey.

Ceredigion Museum, Coliseum, Terrace Road, Aberystwyth, Dyfed. Telephone: Aberystwyth (0970) 617911.
Cliffe Castle, Spring Gardens Lane, Keighley, West Yorkshire BD20 6LH. Telephone: Keighley (0535) 64184.
Clitheroe Castle Museum, Castle Hill, Clitheroe, Lancashire. Telephone: Clitheroe (0200) 24635.
Colne Valley Museum, Cliffe Ash, Golcar, Huddersfield, West Yorkshire. Telephone: Huddersfield (0484) 659762.
Museum of English Rural Life, The University, Whiteknights, Reading, Berkshire RG6 2AG. Telephone: Reading (0734) 875123.
Northampton Central Museum and Art Gallery, Guildhall Road, Northampton NN1 1DP. Telephone: Northampton (0604) 34881 extension 391.
Ordsall Hall Museum, Taylorson Street, Salford, Lancashire M5 3EX. Telephone: 061-872 0251.
Rochdale Museum, Sparrow Hill, Rochdale, Lancashire OL16 1AF. Telephone: Rochdale (0706) 47474 extension 769.
Salford Museum and Art Gallery, Peel Park, The Crescent, Salford, Lancashire M5 4WU. Telephone: 061-736 2649.
Scolton Manor Museum, Scolton, Spittal, Haverfordwest, Dyfed. Telephone: Clarbeston (043 782) 328.
F. Walkley (Clogs) Ltd, Clog Mill, Burnley Road, Hebden Bridge, West Yorkshire. Telephone: Hebden Bridge (042 284) 206. The only surviving clog and sole mill in Britain; visitors may browse around the works and gift shop.
Welsh Folk Museum, St Fagans, Cardiff CF5 6XB. Telephone: Cardiff (0222) 569441.

FURTHER READING

Broomhead, Duncan. *Clogmakers*. Manchester District of the English Folk Dance and Song Society, 1983.
Dobson, Bob. *Concerning Clogs*. Dalesman Publishing Company, 1979.
Edlin, Herbert. *Woodland Crafts in Britain*. David and Charles, 1973.
Fitz Randolph, Helen, and Hay, Doriel. *The Rural Industries of England and Wales*, volumes I and II. EP Publishing Ltd, 1977 (first published 1926).
Hartley, Dorothy. *Made in England*. Eyre Methuen, 1977 (first published 1939).
Jenkins, J. Geraint. *Traditional Country Craftsmen*. Routledge and Kegan Paul, 1965.
Vigeon, Evelyn. 'Clogs or Wooden Soled Shoes'. *Costume* magazine.

CLOGMAKERS

Jeremy Atkinson, Kite Clogs, Capuchin Yard, Church Street, Hereford. Telephone: Hereford (0432) 274269.
Hywel Dafis, Clocsiau Caron, Tregaron, Dyfed.
John the Fish, 2 Tarrandean Bungalows, Perranwell Station, Truro, Cornwall.
Rick Rybicki, Oxford House, Dale Street, Todmorden, Lancashire. Telephone: Todmorden (070 681) 6348.